HOW CAR ENGINE WORKS

Beginners guide on how car engine works,
basic engine components, ignition train
system with how four cylinders and six
cylinder engines differs

Table of Contents

CHAPTER ONE 3

    INTRODUCTION 3

    ENGINE COMPONENTS 4

    INTERNAL BURNING 7

    STROKE OF COMPRESSION 8

CHAPTER TWO 10

    BASIC ENGINE COMPONENTS 10

    ESSENTIAL ENGINE COMPONENTS 12

    ENGINE ISSUES 15

CHAPTER THREE 21

    IGNITION AND ENGINE VALVE TRAIN SYSTEMS.............21

    HOW DOES THE IGNITION SYSTEM WORK.....................22

    SYSTEMS FOR STARTING, AIR INTAKE, AND ENGINE
    COOLING 23

    CHAPTER FOUR 32

    INCREASING ENGINE POWER.........................32

    ENGINE ISSUES AND SOLUTIONS 37

    HOW DO FOUR-CYLINDER AND SIX-CYLINDER ENGINES
    DIFFER 41

THE END 45

CHAPTER ONE

INTRODUCTION

A car is a device that greatly simplifies our transportation needs. It assists it in performing its function by using its own motor engine. More than 10 billion cars are in use worldwide today. The product that is recycled the most globally is cars. Nearly 95% of cars that are no longer roadworthy are recycled. An average car contains about 30,000 unique parts. A car engine may appear to be a complicated maze of metal, wires, and tubes, but it actually powers the vehicle. Because fuel and air inside the engine burn to produce energy, this type of engine is known as an internal combustion engine.

ENGINE COMPONENTS

Engine Block, Combustion chamber, piston, cylinder head, crankshaft, camshaft, timing system, valve train, rocker arms, lifters, fuel injectors, and spark plug

Brief facts:

1. An internal combustion engine produces the small, controlled explosions needed for the car to move. In car engines, a four-stroke design is employed. Intake, exhaust, compression, and combustion are the four strokes.

2. Every day, about 165 automobiles are produced.

3. The cruise control's creator was blind.

4. Internal combustion engines come in a variety of varieties, including diesel and gas turbine engines. Have you ever opened your car's hood and wondered what was happening inside? To the uninitiated, a car engine can appear to be a huge, confusing jumble of metal, tubes, and wires. You might be curious and want to know what's happening. Or perhaps you're in the market for a new vehicle and you hear terms like "start/stop technology," "turbocharged "and" 2.5-liter incline four." What does it all mean? In this book, we'll go over the fundamental concept behind an engine before delving into the specifics of how everything fits together, potential problems, and performance-boosting techniques. A gasoline car engine's

function is to transform fuel into motion so that your car can move. Burning gasoline inside an engine is currently the simplest method for turning a fuel source into motion. Because combustion occurs internally, a car engine is an internal combustion engine. Internal combustion engines come in a variety of varieties. Gas turbine engines are one kind, and diesel engines are another. Each has benefits and drawbacks of its own. The external combustion engine is another option. The best illustration of an external combustion engine is the steam engine found in vintage trains and steamboats. In a steam engine, the fuel coal, wood, or oil burns outside the engine to produce steam, which then generates motion inside the engine an internal combustion engine is much smaller and much more efficient than an external combustion engine.

INTERNAL BURNING

Any reciprocating internal combustion engine operates on the following principle: An incredible amount of energy is released in the form of expanding gas when a small amount of a high-energy-density fuel like gasoline is ignited. That energy can be put to interesting uses. For instance, if you can design a cycle that makes it possible to set off explosions similar to this hundreds of times per minute and if you can effectively channel that energy, you have the basis for a car engine. A four-stroke combustion cycle is used in almost every car with a gasoline engine to turn fuel into motion.

STROKE OF COMPRESSION

1. Burning stroke

2. Rear stroke

A connecting rod connects the piston to the crankshaft. The cannon are "reset" as a result of the crankshaft's rotation. What happens as the engine completes its cycle is as follows:

1. To allow the engine to inhale a cylinder full of air and gasoline, the piston begins at the top, the intake valve opens, and the piston descends. The intake stroke is this.

For this to work, only the tiniest drop of gasoline needs to be mixed into the air.

2. Next, the piston ascends once more in order to compress the fuel air mixture. Explosion power increases with compression.

3. The spark plug releases a spark to light the gasoline when the piston reaches the top of its stroke. The piston descends as a result of the gasoline charge explosion in the cylinder.

4. The exhaust valve opens when the piston reaches the bottom of its stroke, allowing the exhaust to exit the cylinder and exit the vehicle through the tailpipe. The engine now takes in another charge of

gas and air as it prepares for the subsequent cycle. The crankshaft of an engine converts the linear motion of the pistons into rotational motion. We want to turn (rotate) the car's wheels with it, so the rotational motion is convenient. Let's now examine every component that functions as a unit to achieve this, starting with the cylinders.

CHAPTER TWO

BASIC ENGINE COMPONENTS

The cylinder, which houses the piston and moves it up and down inside, is the heart of the engine. Most lawn mowers have single-cylinder engines, but most cars have multiple cylinders four, six and eight cylinders are common. The cylinders in a multi-cylinder engine are typically set up in one of three ways, as depicted in the figures to the left: inline, V, or flat also known as horizontally opposed or boxer. A four-cylinder engine with an inline configuration is the inline four that we mentioned earlier. In terms of smoothness, manufacturing costs, and shape characteristics, various configurations have different benefits and drawbacks. They are better suited for particular vehicles as a result of their benefits and drawbacks. The cylinders are arranged in two banks that are at an angle to one another.

ESSENTIAL ENGINE COMPONENTS

Ignition plug

The spark that ignites the fuel/air mixture and causes combustion is provided by the spark plug. Everything must come together at precisely the right time for the spark to occur. Valves when it's time to let in air, fuel, and exhaust, the intake and exhaust valves open appropriately. It should be noted that the combustion chamber is sealed during compression and combustion because both valves are closed.

Piston

A metal cylinder that moves up and down inside the cylinder is called a piston.

Engine Rings

Between the inner and outer edges of the cylinder and the piston, piston rings offer a sliding seal. The rings have two functions:

1. They keep oil in the sump from leaking into the combustion area, where it would be burned and lost.

2. They prevent the fuel air mixture and exhaust in the combustion chamber from leaking into the sump during compression and combustion. The majority of vehicles that "burn oil" and require a quart to be added every 1,000 miles do so because the engine is old and the seals on the rings have worn out. For piston rings, many contemporary vehicles use more advanced materials. This is one of the factors that contribute to engines lasting longer and requiring fewer oil changes.

Securing rod

The piston is connected to the crankshaft by the connecting rod. It has the ability to rotate at both ends, allowing its angle to change as the crankshaft and piston move.

Crankshaft

Like the crank on a jack-in-the-box, the crankshaft converts the piston's up and down motion into circular motion.

Sump

The crankshaft is enclosed by the sump. It has some oil in it, which gathers at the sump's bottom the oil pan.

We'll then discover the potential engine problems.

ENGINE ISSUES

Car engines can experience a variety of issues, including fuel or battery related issues. You start your car one morning and it turns over but won't start. What might be a problem? Knowing how an engine functions now will help you comprehend the fundamental factors that can prevent it from running. A poor fuel mixture, insufficient compression, or a lack of spark is the three main things that can go wrong. Beyond that, countless unimportant factors could lead to issues, but these are the "big three." A bad fuel mixture can happen in a number of ways:

1. The engine is getting air but no fuel because you are out of gas.

2. There may be fuel but insufficient air because the air intake may be blocked.

3. The fuel system could be supplying the mixture with too much or too little fuel, which would prevent proper combustion.

4. The fuel may contain an impurity that prevents it from burning, such as water in your gas tank.

5. Lack of compression the combustion process won't function as it should if the charge of air and fuel cannot be compressed sufficiently. A lack of compression could happen for the following reasons:

1. Your piston rings are worn, which lets compressed air and fuel leak past the piston.

2. If the intake or exhaust valves are not sealing correctly, compression leaks can also occur.

3. The cylinder has a hole in it. The most typical "hole" in a cylinder is where the cylinder's top, also known as the cylinder head, which houses the valves and spark plug, joins the cylinder itself. Typically, a thin gasket is pressed between the cylinder and cylinder head before they are bolted together to ensure a tight seal. Small holes form between the cylinder and the cylinder head if the gasket fails, and these holes lead to leaks.

Lack of spark: There are a number of reasons why the spark may be absent or insufficient.

1. A weak spark results from a worn-out sparkplug or wire leading to it.

2. If the system that sends a spark down the wire is malfunctioning or the wire is cut or missing, there won't be a spark.

3. The fuel will not ignite properly if the spark occurs either too early or too late in the cycle i.e., if the ignition timing is off.

Many additional issues could arise. For instance:

1. You cannot start the engine by turning the key if the battery is dead.

2. If the crankshaft's bearings, which allow it to rotate freely, are worn out, the engine won't start.

3. If the valves do not open and close at the proper times or at all, the engine cannot run because air cannot enter and exhaust cannot exit.

4. The engine will seize if the oil runs out because the piston cannot move up and down freely in the cylinder. All of these elements are in good working order in a properly operating engine. Although perfection is not necessary for an engine to function, you will probably notice when something isn't quite right. An engine is equipped with a variety of systems to assist in turning fuel into motion. The following sections will focus on the various engine subsystems.

CHAPTER THREE

IGNITION AND ENGINE VALVE TRAIN SYSTEMS

The majority of engine subsystems can be implemented using a variety of technologies, and newer technologies can enhance engine performance. Let's examine each of the various subsystems found in contemporary engines, starting with the valve train. The valves and a device that opens and closes them make up the valve train. A camshaft is the opening and closing mechanism. The majority of contemporary engines have overhead cams. The cams on the shaft either directly or via a very short linkage activate the valves. In earlier engines, the camshaft was housed close to the crankshaft in the sump. The crankshaft and camshaft are connected by a timing belt or timing chain, which keeps the valves and pistons in time. The camshaft is set up to rotate at a speed equal to half that of the crankshaft. Due to the requirement for two camshafts per bank of cylinders in many high-performance engines with four valves per cylinder two for intake and two for exhaust, the term "dual overhead cams" was coined.

HOW DOES THE IGNITION SYSTEM WORK

A high-voltage electrical charge is generated by the ignition system and sent to the spark plugs via ignition wires. The distributor, which is conveniently located under the hood of the majority of cars, receives the charge first. One wire exits the center of the distributor, and four, six, or eight wires come out of it depending on the number of cylinders. Each spark plug receives the charge from these ignition wires. Only one cylinder at a time receives a spark from the distributor due to the timing of the engine. Maximum smoothness is provided by this strategy.

SYSTEMS FOR STARTING, AIR INTAKE, AND ENGINE COOLING

Precise connections between a cooling system and the plumbing the radiator and water pump make up the majority of cars' cooling systems. To cool the cylinders, water flows through the passages surrounding them and then through the radiator. The majority of motorcycles, lawn mowers, and a few cars most notably pre-1999 Volkswagen Beetles have air-cooled engines instead. The fins adorning the outside of each cylinder to help with heat dissipation identify an air-cooled engine. In general, air-cooling shortens engine life and reduces performance while making the engine lighter and hotter. Having learned how and why your engine maintains a cool temperature, But why is proper airflow so crucial? The majority of automobiles are normally aspirated, which means that air enters the cylinders without first passing through an air filter. Modern fuel-efficient engines with high performance are either turbocharged or supercharged, which means that air entering the engine is first pressurized so that more air fuel mixture can be squeezed into each cylinder to increase performance. Boost is the name for the pressurization pressure. A turbocharger compresses the incoming air stream using a small turbine that is attached to the exhaust pipe. The compressor is spun by a supercharger that is directly connected to the engine. The turbocharger boosts the power of smaller engines by using hot exhaust to spin the turbine and compress the air. Therefore, a fuel-efficient four-cylinder engine can produce the same amount of horsepower as a six-cylinder engine while using 10 to 30 percent less fuel. It's great to increase the performance of your engine, but what actually occurs when you turn the key to start it? An electric starter motor and a starter solenoid make up the starting system. The starter motor spins the engine a few times when you turn the ignition key to initiate the combustion process. A strong motor is required to turn a cold engine. The piston rings' total internal friction, which the starter motor must overcome

1. Any cylinders that are currently in the compression stroke and their compression pressure

2. The power required by the camshaft to open and close valves

3. Every other component that is directly connected to the engine, such as the alternator, water pump, and oil pump. A car's 12-volt electrical system can only supply so much power, so hundreds of amps of electricity must go into the starter motor. The starter solenoid, which can manage that much current, is essentially a big electronic switch. The solenoid, which powers the motor, is turned on when you turn the ignition key. The engine subsystems that control what goes in oil and fuel and what comes out will be discussed next exhaust and emissions. Systems for fuel, exhaust, electrical, and engine lubrication the exhaust pipe and the muffler are parts of your cars exhaust system.

Your primary concern when it comes to routine auto maintenance is probably how much gas is in your vehicle. How do the cylinders get their energy from the gas you put in? The fuel system of the engine pumps fuel from the fuel tank and blends it with air so that the right amount of fuel and air can flow into the cylinders. Modern vehicles typically use port fuel injection or direct fuel injection to deliver fuel. In a fuel-injected engine, the appropriate amount of fuel is individually injected into each cylinder, either directly into the cylinder or just above the intake valve port fuel injection direct fuel injection. Older cars were carbureted, which meant that as the air entered the engine, a carburetor mixed the gas and air.

Oil also has a significant role. Every moving component in the engine receives oil thanks to the lubrication system, allowing it to move with ease. Pistons so they can slide easily within their cylinders and any bearings that permit components like the crankshaft and camshaft to freely rotate both require oil The oil pump in the majority of cars draws oil from the oil pan, passes it through the oil filter to remove any grit, and then squirts it at high pressure onto the cylinder walls and bearings. The cycle then repeats as the oil drips into the sump, where it is again collected. Let's look at some of the things that come out of your car now that you are aware of some of the things you put in it. The exhaust pipe and the muffler are parts of the exhaust system. You would hear the sound of tens of thousands of tiny explosions coming from your tailpipe if you didn't have a muffler. A muffler muffles the noise. Modern automobiles' emission control systems are made up of a catalytic converter, a number of sensors and actuators, and a computer that keeps track of everything and makes necessary adjustments. For instance, the catalytic converter burns off any leftover fuel and some other chemicals in the exhaust using a catalyst and oxygen. Making sure there is enough oxygen available for the catalyst to function and making adjustments as needed are done by an oxygen sensor in the exhaust stream.

What else powers your car besides gas? The battery and alternator make up the electrical system. The alternator, which produces electricity to recharge the battery, is belted to the engine. Through the wiring of the car, the battery supplies 12-volt power to all electrically-dependent components such as the radio, headlights, wipers, power windows, power seats, computers, and the ignition system.

CHAPTER FOUR

INCREASING ENGINE POWER

The power and performance of an automobile's engine can be improved by adding a turbocharger. With the help of all of this knowledge, you can start to realize that there are numerous methods for improving an engine's performance. All of the following factors are constantly being tweaked by automakers to increase an engine's output or fuel efficiency. Amplify displacement Because you can burn more gas per engine revolution, more displacement equates to more power. By enlarging the cylinders or by including more cylinders, you can increase displacement. The practical maximum seems to be twelve cylinders. A higher compression ratio Up to a certain point, higher compression ratios result in more power. However, the likelihood of the air fuel mixture igniting spontaneously increases with compression before the spark plug ignites it. Gasoline's with a higher octane rating stop this kind of premature combustion. Because of the higher compression ratios used by their engines to produce more power, high-performance cars typically require high-octane gasoline.

Put more into each cylinder: By packing more air and thus fuel into a cylinder of a certain size, you can increase the cylinder's output much like you would if you increased the cylinder's size without increasing the amount of fuel needed for combustion. The incoming air is pressurized by turbochargers and superchargers in order to efficiently pack more air into a cylinder.

Bring in cooler air: Air's temperature rises when it is compressed. The air in the cylinder should be as cool as it can be, though, since the hotter the air is, the less it will expand when combustion occurs. Consequently, an intercooler is a common feature of cars with turbo and supercharging. A special radiator called an intercooler is used to cool compressed air before it enters a cylinder.

Make it easier for air to enter: Air resistance can drain an engine's power as a piston descends during the intake stroke. Adding two intake valves to each cylinder significantly reduces air resistance. Intake manifolds on some more recent cars are polished to reduce air resistance there. Airflow can also be enhanced by larger air filters.

Make it easier for exhaust to exit: The engine loses power if air resistance makes it difficult for exhaust to exit a cylinder. Each cylinder can have a second exhaust valve added to reduce air resistance. Performance is increased by having four valves per cylinder in a car with two intake and two exhaust valves. When a car advertisement says the vehicle has four cylinders and 16 valves, it actually means that each cylinder of the engine has four valves. Back-pressure can result from an exhaust pipe that is too small or a muffler that has a lot of air resistance, both of which can have the same result. To remove back-pressure in the exhaust system, high-performance exhaust systems use headers, large tail pipes, and free-flowing mufflers. When a car is described as having "dual exhaust," it means that there are two exhaust pipes instead of one, which improves the flow of exhaust.

Lighten everything up: The performance of the engine is improved by lighter parts.

A piston must expend energy each time it changes direction in order to stop moving in one direction and begin moving in another. The piston requires less energy the lighter it is. As a result, both performance and fuel efficiency improve.

Put the fuel in: Fuel injection enables very accurate fuel metering to every cylinder. Performance and fuel efficiency are improved.

ENGINE ISSUES AND SOLUTIONS

1. What distinguishes a gasoline engine from a diesel engine? There is no spark plug in a diesel engine. Instead, diesel fuel is injected into the cylinder, where it ignites

due to the heat and pressure of the compression stroke. Diesel engines get better mileage because diesel fuel has a higher energy density than gasoline.

2. What distinguishes a two-stroke engine from a four-stroke engine? Two-stroke engines are the norm for chainsaws and boat motors. Because there are no moving valves in a two-stroke engine, the spark plug ignites each time the piston reaches the top of its cycle. Gas and air enter the cylinder through a hole in the lower portion of the wall. Combustion is started by the spark plug as the piston rises, and exhaust leaves the cylinder through a different hole. In a two-stroke engine, you can't use rings to seal the combustion chamber

because of the holes in the cylinder wall, so you have to mix oil into the gas. A two-stroke engine typically generates a lot of power relative to its size because there are twice as many combustion cycles per rotation. A two-stroke engine is much more polluting because it consumes more gasoline and burns a lot of oil.

3. Are there any advantages between steam engines and other external combustion engines, as I mentioned in this book? The main benefit of using anything that burns as fuel for a steam engine is this. For instance, whereas an internal combustion engine requires pure, high-quality liquid or gaseous fuel, a steam

engine can run on coal, newspaper, or wood as fuel.

4. Why does an engine need eight cylinders? Why not replace the eight cylinders with a single large cylinder with the same displacement? Eight half-liter cylinders rather than a single large 4-liter cylinder are used in a large 4.0-liter engine for a few different reasons. The smoothness is the primary factor. Because there are eight evenly spaced explosions in a V-8 engine rather than one large explosion, the engine is much smoother. Starting torque is an additional factor. When a V-8 engine is started, only two cylinders 1 liter are driven through their compression strokes; however, if there

were only one large cylinder, four liters would need to be compressed.

HOW DO FOUR-CYLINDER AND SIX-CYLINDER ENGINES DIFFER

With 325 horsepower and 380 lb.-ft. of torque, the 2.7-liter Eco Boost engine that powers the 2017 Fusion V6 Sport is standard ford. An engine's overall performance is significantly impacted by the number of cylinders it has. The pistons inside each cylinder pump air into the cylinder, connecting to the crankshaft and turning it. More combustive events are occurring at any given time the more pistons are pumping. Thus, more power can be produced in a shorter amount of time. While 6-cylinder engines are frequently set up in the more compact "V" shape, which is why they are known as V6 engines, 4-cylinder engines frequently come in "straight" or "inline" configurations. V6 engines were the preferred option for American automakers due to their strength and quietness, but turbo charging technologies have increased the power and appeal of four-cylinder engines to consumers. Four-cylinder engines were once despised by American auto buyers who thought they were weak, unbalanced, slow to accelerate, and slow. But in the 1980s and 1990s, when Japanese automakers like Honda and Toyota started putting highly efficient four-cylinder engines in their vehicles, Americans developed a new appreciation for the small engine. Japanese models, like the Toyota Camry, started outselling American models quickly. Modern four-cylinder engines, like Ford's Eco Boost engine, use lighter components and turbo charging technology to achieve V-6 performance from more efficient four-cylinder engines. These smaller turbocharged engines are less stressed thanks to cutting-edge aerodynamics and technologies, In terms of the V6's future, the gap between four-cylinder and V6 engines has significantly narrowed recently. V-6 engines are still useful, though, and not just in sports cars. The power of a V-6 is required for trucks that are used to haul loads or tow trailers.

In those situations, strength takes precedence over effectiveness.

THE END

www.ingramcontent.com/pod-product-compliance
Lightning Source LLC
Chambersburg PA
CBHW071116220526
45467CB00004B/1913